AF887693

Leaping Learners
Education, LLC

For more information and resources visit us at:
www.leapinglearnersed.com

©Sean Bulger

Published by Leaping Learners, LLC, Verona, NJ 07044 . All rights reserved. Reproduction in whole or in-part without written permission of the publisher is prohibited.

Every attempt has been made to credit each photo. Please contact us if there has been and error and we will resolve the issue.

Cat vector ©Andrey1005/shutterstock.com, Cover © Ewa Studio/shutterstock.com, Pg 1 © acceptphoto/shutterstock.com, Pg 2 © PickOne/shutterstock.com,Pg 3 © ramarama/shutterstock.com,Pg. 4 © LittlePerfectStoc k/shutterstock.com,Pg. 5 © islavicek/shutterstock.com, Pg. 6 © Alexander Raths,/shutterstock.com, Pg. 7 © Bufo/shutterstock.com, Pg. 8 © Dmitry Kalinovsky/shutterstock.com,pg. 9 © Mr.Samarn Plubkilang/shutterstock.com, Pg. 10 © SGr/shutterstock.com, pg. 11 © Aumsama/shutterstock.com,

All design by Sean Bulger

ISBN
978-1-948569-28-6

Dear Parents and Guardians,

Thank you for purchasing a *Katie the Cat* series book! After teaching students from kindergarten to second grade for over seven years, I became frustrated at the lack of engaging books that my students could read independently. To help my students engage with nonfiction topics, my wife and I decided to write non-fiction books for children. We hope to excite young children about the natural world.

Here at Leaping Learners, LLC we have 3 main goals:

1. Spark young readers' curiosity of the natural world
2. Develop critical independent reading skills at an early age
3. Develop reading comprehension skills before and after reading

We hope that your child enjoys learning with the *Katie the Cat.* If you or your children are interested in a topic that we have not written about yet send us an email with your topic, and maybe your book will be next!

Thank you,

Sean Bulger Ed.M

Reading Suggestions:

Before reading this book, encourage your children to do a "picture walk" where they skim through the book looking the pictures to help them think about what they already know about the topic. Thinking back about what they already know helps children to get excited about learning more facts and begin reading with some confidence.

After previewing the book, encourage your child to read the book independently more than once. After they have read it, ask them specific questions related to the information in the book.

Finally, see if your child can complete the reading comprehension exercises at the end of the book!

This book is best for ages 4-6 but......
Please be mindful that reading levels are a guide and vary depending on a child's skills and needs.

Katie Teaches You About. . . Pigs

Written by Sean and Anicia Bulger

This is a pig.

These are
baby pigs.

Pigs have a nose called a **snout**.

say...

sn - out

snout

Pigs have four legs.

Pigs have
feet called
hooves.

Some pigs are big.

Pigs eat apples.

Some pigs
have pink skin.

Pigs eat food together.

Pigs live outside on farms.

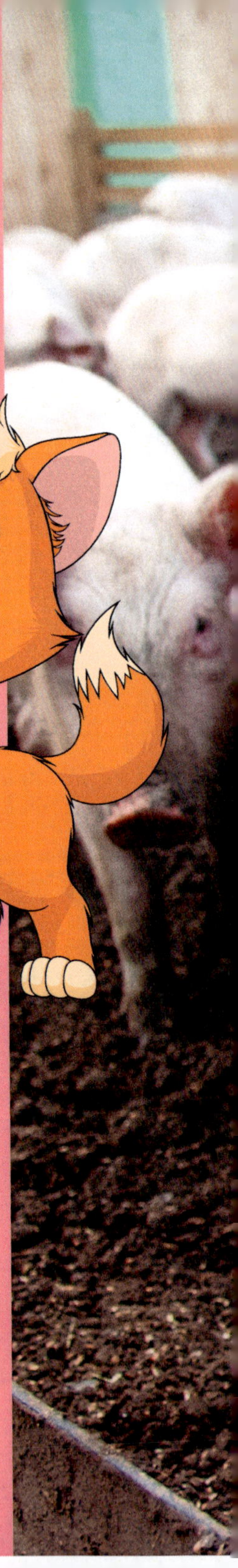

Pigs like the mud.

Draw a pig.

This is a ________________.

Draw a snout.

This is a ______________.

Find and circle the word "mud"

mud

have

they

grass

has

The

they

mud

am

has

mud

we

grass

Find and circle the word "farm"

girl

am

farm

See

get

get

See

We

See

water

farm

farm

get

am

girl

you

Want to learn about colors? Check out the "Clayton Teaches you About..." Series!

Great for early readers ages 4-6

Want to learn about rainforest animals? Check out the "Matt Learns About..." series!

Great for emerging readers ages 6-8

Want to learn about ocean animals? Check out the "Fay Learns About..." series!

Great for emerging readers ages 6-8

Made in United States
Orlando, FL
02 November 2026

90104233R00021